FARMING

FARMING

Growing the Food that Feeds Us

CHRIS McNAB

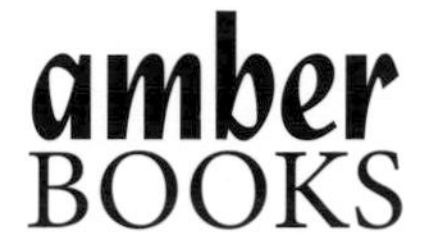

Published by Amber Books Ltd
United House
North Road
London N7 9DP
United Kingdom
www.amberbooks.co.uk
Instagram: amberbooksltd
Facebook: amberbooks
Twitter: @amberbooks
Pinterest: amberbooksltd

ISBN: 978-1-83886-255-8

Project Editor: Michael Spilling
Designer: Keren Harragan
Picture Research: Terry Forshaw

Printed in China

Contents

Introduction

For the many of us who don't farm the land, it's easy to forget the fact that all human life depends upon soil and the plants that grow within it. The agricultural revolution that occurred about 12,000 years ago unchained humanity from the precarious existence of the hunter-gatherer. From this point, people could stay in one place and farm seasonal crops and domesticated livestock. Population growth, urbanization and civilization itself were the results. The word 'farming', however, covers a vast catalogue of practices, crops, animals and industries, from avocadoes to wheat, potatoes to pomegranates, cattle to llamas. Then there are the differences of scale: today, more than 2 billion people on the planet still work small plots of land intimately as subsistence farmers, often with tool types dating back millennia, but agriculture is also found in multi-million dollar industries, with farms covering tens of thousands of acres. While recognizing our potentially fragile dependency on the earth beneath our feet and those who work it, this book is a celebration of all forms of agriculture, large and small, across our fertile planet.

ABOVE:
Soybean harvesting, Ontario, Canada

OPPOSITE:
Stepped rice padi, Bali, Indonesia

Europe

Europe, despite the dense urbanization of many of its countries, remains an intensely agricultural continent. One of the keynotes of European agriculture is its diversity of crop production, courtesy of the extremely wide variations in soil types and seasonal climates. The United Kingdom (UK), for example, has about 70 per cent of its landmass given to agricultural practices, of which c. 39 per cent is arable land. From rich soils and a temperate climate, the UK's main arable products are (in descending order of volume) wheat, potatoes, rapeseed, sugar beet, barley, carrots, turnips, strawberries, apples, onions and lettuce.

Spain has a similar percentage of land available for cultivation, but its growing conditions are very different from those of the UK, with generally lower levels of soil quality and significantly less annual rainfall, higher overall temperatures, more sunlight and less cloud cover during the summer months. The country's distinctive climatic conditions make it one of the world's biggest producers of olives, grapes, tomatoes, oranges, tangerines, chilli peppers, artichokes and lemons. And these are just two of 44 countries in Europe. Within European boundaries, therefore, the spectrum of farm types, farming methods and agricultural outputs is impressive, although only approximately 4 per cent of Europe's working population is involved in farming. Compared to countries such as the United States (USA), however, European farms are relatively small – the average farm in the EU-28 in 2016 had just 16 hectares (39.5 acres) of land.

OPPOSITE:
Dairy farming, Cantabria, Spain
Asturian Mountain cattle rest in an idyllic mountain location. The breed is reared for its beef and also for its milk, which is used to produce the famed regional Casín cheese.

RIGHT:
Hand milking, Poland
Hand milking is a practice as ancient as domesticated cattle. Hand milking an individual cow takes roughly 10 minutes, depending on the size of the animal. The farmer works from the front quarters to the back quarters.

OPPOSITE LEFT:
Dairy farming, Oderberg, Germany
One of the most significant automations in livestock farming during the 20th century was the automatic milking machine (AMM), which enabled the dramatic upscaling of milk outputs from dairy herds.

OPPOSITE RIGHT:
Dairy farming, Oderburg, Germany
Although a multi-cow AMS might appear an alien environment for the cow, it can deliver a more consistent milking experience for the animal, reducing the creature's stress levels as it becomes more familiar with the process.

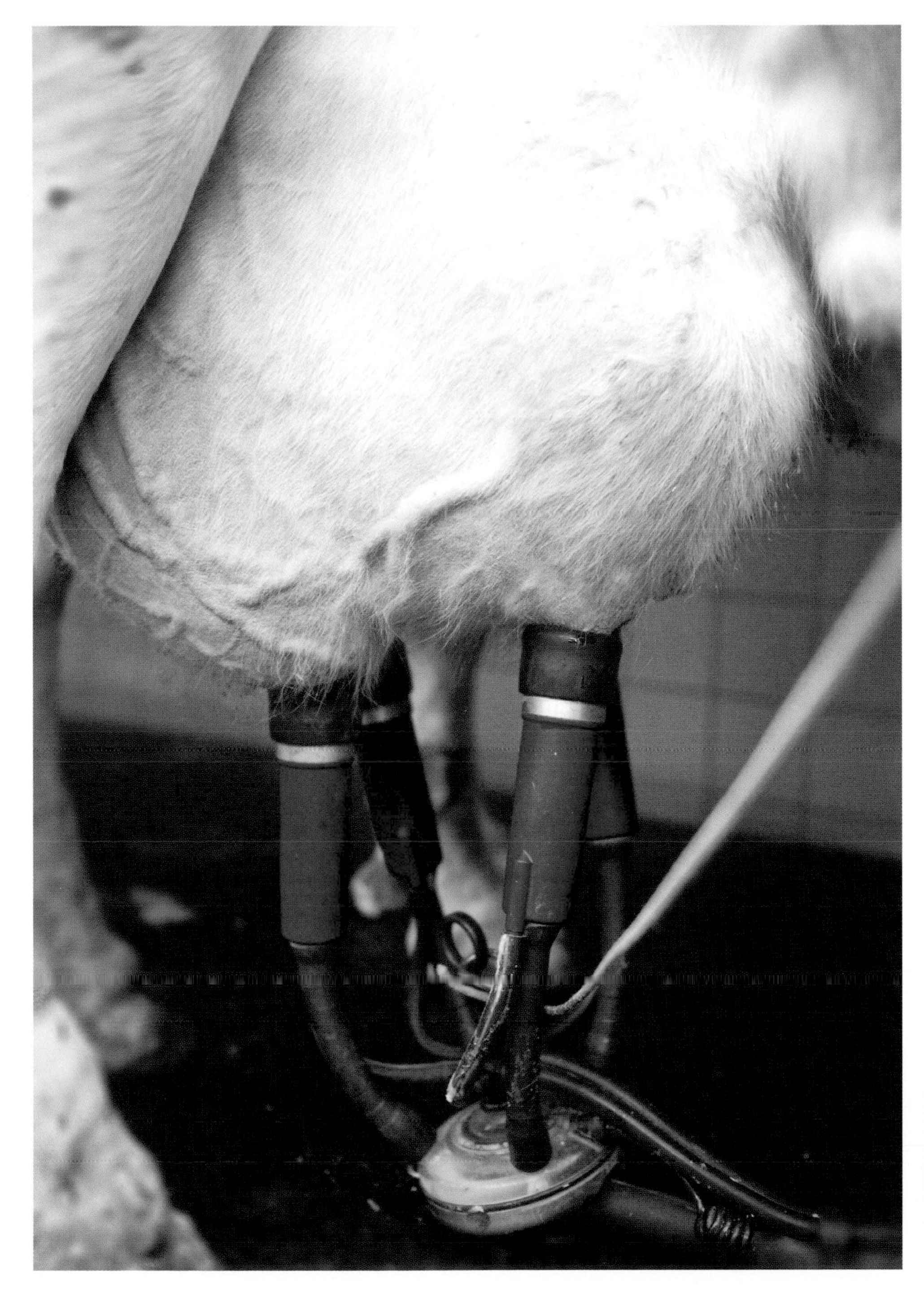

46 643
46 643

LEFT:
Cattle farming, Iceland
Cattle farming is one of the main agricultural activities in Iceland, which has about 74,500 cattle, approximately 26,000 of which are dairy cattle. All of these calves are fitted with ear tags, on which there is information such as the gender, age and weight of the calf and the identity of the calf's mother.

ABOVE:
Grass harvesting, Shropshire, England, UK
Using a tractor and a forage harvester machine, a British farmer collects grass that will be used in the production of silage. Silage is grasses or other green foliage crops that have been fermented in a low-oxygen environment, such as being tightly packed under weighted plastic sheeting. The eventual product is used as fodder for animals.

LEFT:

Cattle farming, Liechtenstein

About 16 per cent of Liechtenstein's territory consists of highland pasture for a small livestock industry. The cattle will graze these meadows during the summer months, being taken to lowland areas before the winter sets in.

OPPOSITE:

Cattle farming, Hungary

The Hungarian Grey is a particularly impressive breed of beef cattle, defined most conspicuously by its elongated and often lyre-shaped horns. The animal is also very strong, which historically has meant that it has been used extensively as a draught animal.

BAGSHAWS

LEFT:

Bagshaws livestock market, Derbyshire, England, UK

Bagshaws livestock auctioneer was established in 1871 and grew to occupy several auction sites across the UK, including this one at Bakewell in Derbyshire. Livestock auctions are big business – in 2017 alone, 1.4 million cattle and 10 million sheep were traded in auctions in England and Wales.

RIGHT:

Bull, Sweden

For cattle farmers, breeding bulls is an expensive but crucial herd investment. Superior genetic material in the bull can mean easier, more successful gestations for the cows and healthier, heavier calves at birth.

LEFT TOP:

Deer farming

Deer are farmed for deer meat (venison) but also for several important by-products: hides, velvet, antlers and musk. Alongside the commercial aspects of deer farming, however, the industry is also used to breed and conserve rare or endangered species of deer.

LEFT BOTTOM:

Domestic geese

Goose farming can be a more profitable enterprise than duck farming, as geese live off simple and abundant grasses and grow far larger than both ducks and chickens.

RIGHT:

Family pigs, Forest d'Aïtone, Corsica, France

These free-ranging pigs are fortunate enough to have escaped intensive farming methods. For the owners of a smallholding, pigs require intensive year-round investment in time, food and fencing, but a mature pig will provide plentiful pork, bacon and sausages.

FAR LEFT:

Salmon farming, Loch Awe, Argyll and Bute, Scotland, UK

An open-net pen salmon farm in Scotland, seen from the air. Open-net pens are controversial because surrounding open water flows freely through them, allowing fish waste and potential diseases to pass from the pen into the wider ecosystem.

LEFT:

Salmon farming, Loch Awe, Argyll and Bute, Scotland, UK

Salmon farming is, historically speaking, a relatively new industry, having been established in the 1980s. Today, however, about 60 per cent of the world's salmon is produced in farms rather than being caught in the wild.

Sheep farming, Capel Curig, Conwy, Wales, UK
The small nation of Wales in the UK is known globally for its sheep farming, which produces some of the world's finest lamb and wool. Wales has more than 10 million sheep, a figure that far exceeds the country's human population of c. 3 million. Most of the sheep farms are located on island mountains and moorlands or along the south and west coast.

ABOVE:
Sheep, Moel Siabod, Snowdonia, Wales, UK
Wales' mountainous terrain is exposed to extremes of climate, thus the sheep breeds farmed there are necessarily hardy. The Welsh Mountain Sheep is one of the oldest sheep breeds in history, with references to the breed in medieval literature.

RIGHT:
Hill farming, Llangattock Mountain, Powys, Wales, UK
Sheep farming in mountainous areas is a tough business, and UK hill farmers often struggle to make a profit, despite subsidies towards their roles in wilderness maintenance. Losses to predators, theft and weather can be high and herding requires expert use of vehicles and dogs.

RIGHT:
Sheep, Wales, UK
In sheep farming for meat, lambs tend to stay with their mothers for the first 8–14 weeks of life, with the ideal weaning age being 12–14 weeks. If the sheep are used for their milk, however, the lambs might be removed from their mother 24–36 hours after lambing, and fed on a milk substitute.

FAR RIGHT:
Farmhouse, Nant Ffrancon, Snowdonia, Wales, UK
Wales is dotted with farmhouses situated in the most remote, rugged and beautiful locations. Many of the older farmhouses have been in a single family for many generations.

RIGHT:
Sheep shearing, UK
Sheep shearing is a back-breaking process requiring both speed, skill and strength on the part of the shearer. The 2021 world record was 872 strong wool lambs sheared in nine hours, set by Oxfordshire shearer Stuart Connor.

FAR RIGHT:
Royal Welsh Show, Llanelwedd, Wales, 2013
The Royal Welsh Show is one of the largest agricultural events in the British farming calendar, held annually over a four-day period. It includes huge livestock displays, such as this one of sheep breeds.

Welsh Half-Bred
Defaid Rhan-Frid Cymru
Welsh Mule
Defaid Miwl Cymreig
Beulah Speckled Face
Defaid Penfrith Beulah
Disgarth Isaf
BLAENBLODAU FLOCK
GLYNGLAS BALWENS
GWYNION MAW

OPPOSITE:

Sheep grazing, Col du Tourmalet, Pyrenees mountains, France

Comparatively, France is a small producer of lamb, ranking 29th globally for the years 1961 to 2019. Much of its sheep farming remains traditional rather than industrial in type, especially in the mountainous regions.

LEFT:

Sheep flock, Provence, France

In a picturesque scene, a flock of sheep is driven through the narrow streets of small villages in Provence. The practice of walking livestock over distance is known in English as 'droving'.

Vineyards, Marne Valley, Champagne, France
The historical province of Champagne in north-east France famously produces the sparkling wine that carries its name. The Vallée de la Marne is the largest of the five wine-growing districts in Champagne, with some 11,500 hectares (28,417 acres) of vineyards.

OPPOSITE:

Olive grove, Greece
Olive cultivation has been practised in Greece for at least 4,000 years, the olive tree referred to by the ancient playwright Sophocles as 'the tree that feeds our people'. Olives thrive in hot, sunny and sheltered locations.

FAR LEFT TOP:

Harvested olives, Italy
Olives are harvested from August to November. Their colour varies from green and yellow, through red and purple to black, depending on their ripeness.

FAR LEFT BOTTOM:

Olive oil production, Abruzzo, Italy
The first pressing of olive oil is collected at an *oleificio* in the Abruzzo region of Italy. The olives chosen for oil production tend to be of light yellow colour; the oil content of the olives is reduced as they ripen further.

LEFT:

Olive oil bottling, Jaén province, Andalusia, Spain
Olive oil production requires a large and successful harvest. It typically takes 36–45 kg (80–100 lb) of olives to produce 3.8 litres (1 gallon) of olive oil.

OPPOSITE:

Vineyard, Tuscany
Tuscany is one of the great wine-producing regions of Italy. The Sangiovese grape is the dominant variety here, accounting for 85 per cent of the red wine volume, the wines including Chianti, Brunello di Montalcino and Vino Nobile di Montepulciano.

LEFT TOP:

Grapes, Bordeaux, France
Although wine grapes clearly resemble table grapes, there are key differences. Wine grapes are much smaller than table grapes, have a higher seed content, higher levels of natural sugars, thicker skins and a greater proportion of juice to pulp.

LEFT BOTTOM AND ABOVE:

Château Margaux, Bordeaux, France
The Château Margaux estate in Bordeaux produces some of the finest, and most expensive, French wines. Here, we see barrels of wine elegantly stored in a *chai* (wine storeroom) and undergoing mixing.

RIGHT:

Grape harvesting, Germany

For centuries, grape harvesting been performed by hand, but some vineyards have adopted mechanical harvesting machines. These devices operate by shaking the grapes off the plant through contact rods and collecting them as they fall on a belt that wraps around the stem of the vine.

FAR RIGHT:

Mosel Valley, Germany

The Mosel River glides through a valley heavily cultivated with vineyards. Wine producing in the Mosel region dates back to the 2nd century BCE and today the vineyards cover 9,034 hectares (22,320 acres), much of this area given over to the Riesling grape.

ALL PHOTOGRAPHS:

Wine production, Elciego, Rioja Alavesa, Spain

The Marqués de Riscal winery was founded in 1858 and has grown to be one of the most important wineries in the Rioja, wine region, exporting its wines to more than 110 countries globally. The Rioja wine region lies in northern Spain, and is separated into three sub regions: Rioja Alta, Rioja Alavesa and Rioja Oriental.

13

E3216
BFH

LEFT:
Grape harvesting, Kent, England, UK
Most grape harvesting is still performed by hand, as this process results in less damage to the fruit than automated processes, plus the pickers can do quality inspection as they work.

OPPOSITE LEFT TOP:
Grapes, Kent, England, UK
England's wine growing industry is concentrated around the southern and south-eastern coastal regions, the three major counties being Sussex, Kent and Surrey, where the climate is slightly warmer and drier than elsewhere.

OPPOSITE LEFT BOTTOM:
Plums, Kent, England, UK
Ripe plums hang ready for picking in a traditional orchard in Kent. Plum trees thrive in moist but well-drained soil and under plenty of sunlight.

OPPOSITE RIGHT:
Apple harvesting, Kent, England, UK
English Cox apples are harvested in an orchard in Kent. The English Cox is one of the defining British apples, with a rosy skin and crisp, sweet white flesh, although it has been overtaken in popularity by Gala and Braeburn, which are heavier-cropping varieties.

RIGHT:
Tangerines, Mallorca, Spain
The Balearic island of Mallorca has a highly productive citrus fruit industry, mainly concentrated around the western coastal town of Sóller.

OPPOSITE LEFT:
Orange groves, Spain
An aerial view of these Spanish orange groves reveals charming geometric patterns. Spain is one of the world's largest orange producers, with an output of 3.3 million tonnes (3.6 million US tons) in 2020.

OPPOSITE RIGHT TOP:
Apricots, Spain
A Spanish tree hangs heavy with ripe apricots. With the diversification of diets in the developed world, apricots have become a more profitable crop for many fruit farmers.

OPPOSITE RIGHT BOTTOM:
Lemon harvesting, Spain
The world's leading lemon producers are Spain, Mexico, the Netherlands, South Africa and Turkey, with the major export markets being Europe and the USA.

RIGHT:

Organic farm, London, UK

This aerial photograph shows a small organic farm developed on a patch of land in London. In a 2013 report, the United Nations strongly advocated a return to small-scale local farming as a solution to the growing global food crisis.

OPPOSITE:

Polytunnels

These are elongated semicircular tunnels made out of polyethylene. In agriculture, they are used to create an internal microclimate that is hotter and more humid than the outside, thereby creating the conditions to grow certain fruit and vegetables even out of season.

OPPOSITE:

Bio-farm, the Netherlands

This bio-farm in the Netherlands is cultivating vegetables through a hydroponic method. Hydroponics refers to growing plants without the use of soil, replacing the soil with a nutrient-rich water solvent. Facilities such as this are able to grow seasonal crops throughout the year.

LEFT:

Tomato growing

Compared to the outside world, greenhouses offer many advantages for growing tender crops such as tomatoes: consistent temperatures, protection from excessive cold or damp, more steady plant growth and more manageable pest control. The main downsides are the expense of installation and maintenance.

RIGHT:

Strawberry growing, Netherlands

Strawberry production in the Netherlands has increased significantly in recent years, with a record 78,000 tonnes (85,980 US tons) grown in 2020 alone. Polytunnels such as these extend the growing season significantly.

OPPOSITE:

Strawberries, Netherlands

Through the application of greenhouses and plastic sheeting, home-grown strawberries are available in the Netherlands as early as April and as late as November. Consumption of strawberries in the country increased by 30 per cent between 2017 and 2020.

LEFT:
Strawberry growing, Netherlands
This Dutch strawberry greenhouse shows strawberries in various stages of ripeness. The installation of controllable LED lighting in the hydroponic greenhouses has been a key factor in extending the growing season into winter.

RIGHT:
Grain harvesting, Moldova
A tractor pulls away from a combine harvester, the trailer full of grain. Some of the world's most capable combines can harvest about 12 hectares (30 acres) of grain crops every hour, although much depends on external factors such as weather, crop height and gradients.

OPPOSITE:

Combine harvester, Germany

The business end of a combine. The wheat is cut at the base by a cutter bar and the reel (the rotary cage-like device) pushes the crop down on to the cutter with horizontal bars called bats.

RIGHT:

Round hay bales, Swinbrook, England, UK

Like pieces on a board game, neat round hay bales lie around this freshly cut field in the Cotswolds. Rolling hay results in the bale being more densely packed than it would be in a traditional square bale, meaning that it is more resistent to moisture penetration.

RIGHT:
Potato harvest, Mutterstadt, Germany
Potatoes are one of the most important crop types grown across Europe. The top five of the European growers (in descending order) are: Germany, Poland, France, the Netherlands and the UK.

OPPOSITE LEFT:
Carrots, Netherlands
Carrots need to be planted in loose, deep and well-drained soils in order to flourish. To achieve this, the fields are ploughed to depths of about 20–30 cm (8–12 in), before being harrowed, levelled and cleaned – it is important to remove as much debris, clods and stones as possible.

OPPOSITE RIGHT:
Organic parsnips, Bristol, England, UK
Organic parsnips are displayed for sale, still coloured by the mud from the fields. Parsnips grow best in open fields that attract plenty of direct sunlight, and in deep and light soil.

OPPOSITE:

Apple orchard, Herefordshire, England, UK

An English apple orchard is in full bloom in the spring. While a garden apple tree might produce about 100–150 apples each season, intensive commercial apple tree varieties might produce 400–800 apples, sometimes even more.

LEFT:

Apple harvesting, Herefordshire, England, UK

Apple harvesting takes place in a very narrow window of time, typically about 8–12 days. Because the fruit is so sensitive to bruising, it is generally harvested by hand.

ALL PHOTOGRAPHS:

Hops farming

Hops are one of the principal ingredients in beer production. The distinctive green flowers (above) are part of a climbing plant, hence hops are often grown commercially upon strings and trellises (right). Europe is dotted with many historic examples of oast houses (opposite) – buildings designed for drying hops during the beer-making process.

Asia and the Pacific

The Asia-Pacific region is our planet's largest agricultural producer, a status achieved not only by prodigious growth in population, affluence, export markets and mechanization, but also by the fact that many of its population still remain wedded to subsistence farming for their livelihood. (Agriculture employs a fifth of the region's workforce.) China and India, with their huge populations and vast expanses of land, have been the engine room of this growth since the 1990s, but most Asian-Pacific countries have experienced their own significant growth patterns. Covering about half of the planet's surface, the region produces a dizzying array of crops and meats, including some of the staple products of global consumerism: rice, tea, sugar, sheep, milk and rubber.

Like many regions of the world, however, Asia and the Pacific are experiencing some severe pressures upon their agricultural systems and future way of life, through a combination of enviromental, economic and social factors. Indonesia, for example, grew 180,000 tonnes (198,400 US tons) of cacao in 2020–21, which sounds impressive until compared against the 410,000 tonnes (452,000 US tons) produced in 2012–13. Overfishing is depleting many freshwater and saltwater stocks, potentially to a point of no return. New Zealand's sheep farming, though still large in scale, is only 40 per cent of what it was in the early 1980s. Climate change is adding its own pressures, and the next 50 years of Asian-Pacific farming are likely to be far tougher than the previous 50.

OPPOSITE:
Rice field, Thailand
Although rice is the salient crop of the Asia-Pacific region, much rice growing still takes place at a small scale using traditional methods. Here, a buffalo is used to pull a plough through a rice field in rural Thailand. Although crude, livestock ploughing offers some economic and environmental advantages compared to machinery.

Rice plantation, Bali, Indonesia
Rice is often grown in terraced plantations, the individual terraces working to decrease water run-off (rice requires heavy and constant irrigation) and soil erosion. Terraces also make it viable to farm steep, hilly land.

RIGHT:
Rice growing, Yuanyang County, China
Millions of people across Asia depend upon rice growing for basic subsistence. Rice requires flooded irrigation to grow successfully, with the water level about 5–10 cm (2–4 in) above the ground at all times – drought is disastrous to a rice crop.

FAR RIGHT:
Rice planting, Yogyakarta, Indonesia
Much of global rice farming is still performed by hand. To intensify the yields from their land, the farmers often plant the rice on dry land or in a nursery, then transfer the seeding to the flooded fields for growth to harvest.

OPPOSITE:

Rice harvest, Thailand

These Thai farmers are engaged in the process of winnowing their organic rice harvest. Winnowing involves separating the quality rice grains from the chaff. By throwing the rice into the air, the heavier grains fall to the ground while the chaff is blown away on the breeze created by the man wielding the fans.

LEFT:

Rice straw packing, Hebei Province, China

Rice straw is a by-product of rice harvesting, and a billion tons (1.1 billion US tons) of it is produced every year. While it can in theory be collected and repurposed for fertilizating soil, only those farms with the required mechanization can effectively undertake this process.

FAR LEFT:
Agriculture in Ladakh, India
The high-altitude desert region of Ladakh, India, is cold and arid, hence agriculture only thrives in sheltered plains irrigated by the waters from mountain rivers – rivers that in many places are in retreat.

LEFT TOP:
Ploughing, India
The practice of ploughing fields with two yoked oxen dates back to prehistory in India, but remains a common farming method among subsistence farmers to this day. Most ploughshares in India are metal, but traditional wooden ones are also used.

LEFT BOTTOM:
Transporting fertilizer, Nepal
Many of the world's poorest regions still draw upon child labour in farming. Here, two very young children in Nepal climb steep slopes with baskets containing oxen dung, to be used to fertilize upland terraces.

LEFT:
Irrigation, Bali, Indonesia
The *subak* system of Bali is a method of rice paddy field irrigation that incorporates cultural, religious and environmental outlooks – a system in tune with both people and place.

RIGHT:
Sorghum harvesting, Heihe, Heilongjiang, China
A combine harvester digests a crop of sorghum in China. China also imports large volumes of sorghum from the USA, mainly for use as a livestock feed.

OPPOSITE TOP:
Broom-making, Luannan, China
Agriculture is not just about food and nutrition. The by-products of agriculture have numerous purposes, such as the grass stalks here being reworked into broom bristles.

OPPOSITE BOTTOM:
Making sorghum flour, Yingbeigou, China
Sorghum is a flexible cereal crop. It can be cooked as a grain in the same way as rice, popped like popcorn, added to soups or salads or, as seen here, milled into a flour.

61040

LEFT:

Soybean planting, Akola, India

A family of Indian farmers is sowing soybean seeds in the traditional way, using a pair of oxen to pull a basic plough. In recent years, India's soybean output has plunged owing to erratic rainfall patterns.

RIGHT TOP:

Raisin production, Afghanistan

An Afghan farmer tips a crop of grapes on the ground for air-drying into raisins. To make about 0.5 kg (1 lb) of raisins requires roughly 1.8 kg (4 lb) of grapes. The transition takes about two to four weeks.

RIGHT BOTTOM:

Pomegranates, Kandahar, Afghanistan

This Afghan farmer is sorting pomegranates following harvesting. Afghanistan has traditionally been regarded as one of the world's largest and finest producers of quality pomegranates, although conflict has constrained the industry since the 2000s.

LEFT:

Vertical farming, Depok, West Java, Indonesia

This hi-tech farm in Indonesia utilizes the practice of 'vertical farming', in which crops are grown in vertically stacked layers in a controlled environment. The principal benefit of vertical farming is that it can maximize the crop yield from a relatively small area of ground.

OPPOSITE:

Growing organic vegetables, Singapore

An ingenious use of PVC pipes enables organic vegetables to be grown in an urban farm on the rooftop of a multi-storey car park in a public housing estate in western Singapore, 2018.

LEFT:
Fish farm, Fujian, China
Fish farming is a vast industry around coastal China. Two-thirds of the world's aquaculture (the farming of fish in ponds, lakes and man-made environments) takes place in China, although the practice has had a heavy environmental impact. The fishermen tend to live in huts directly above the waters they farm.

ABOVE:
Fish farm, Vietnam
Small-scale fish farming in Vietnam and across wider Asia often takes place in simple wooden cages, several cages being interlinked to form a raft-like structure. (Larger industrial farms tend to use floating, circular, high-density polyethylene net cages.) Marine creatures raised in these cages include catfish, snakehead fish, perch, prawns, seabass and grouper.

RIGHT:
Fish farm, West Benghal, India
These rural villagers are engaged in the farming of hybrid magur (a type of catfish). Here, workers are transferring live fishes from a muddy pond to a netted enclosure.

FAR RIGHT:
Fish farm, Thailand
Thai aquaculture has attracted much controversy over recent years, particularly for overcrowding in fish pens and poor animal welfare in the processing of the fish from pen to table. The fish are often sedated with clove oil before harvesting.

FAR LEFT:

Tea plantation, Assam, India

Rural women workers harvest tender tea shoots on a plantation (or 'tea-garden') in Assam. The three main tea-growing regions in India are Darjeeling (north-east), Assam (north-east) and the Nilgiri Hills (south-east).

LEFT:

Tea harvesting, Darjeeling, India

Workers harvest Darjeeling tea, collecting the fresh leaves in baskets. Darjeeling is one of the finest world teas, and India produces more than 10,000 tonnes (11,000 US tons) of this crop every year.

LEFT:

Tea factory, Munnar, India

In this tea factory in Munnar, the leaves are undergoing the process known as rolling. The rollers on the machine break up leaves that have just come from the withering (dehydration) process. After this stage, the tea will pass on to oxidation (which determines the tea's strength, colour and taste) and drying/firing, prior to packing.

RIGHT:

Tea plantation, Cameron Highlands, Malaysia

An aerial view of a tea plantation in the majestic Cameron Highlands, Malaysia. Industrial tea production in the Cameron Highlands was established by the British in the 1920s, and today the beauty of the plantations makes them a major tourist destination.

ABOVE:

Tea harvesting, Cameron Highlands, Malaysia

This tea plantation worker is collecting tea leaves using a manually operated tea plucking shear, which both strips and collects the leaves. There are powered versions of this device (electric or internal combustion) that dramatically speed the process of harvesting.

RIGHT:

Tea plantation, Sri Lanka

The island of Sri Lanka is home to several major Ceylon tea-growing regions. Many of its tea plantations are found in high-growing (above 1,828 m/6,000 ft) or mid-growing areas (609–1,219 m/2,000–4,000 ft), and often feature precipitous slopes and lofty elevations.

OPPOSITE:

Cacao harvesting, Gorontalo, Indonesia

Indonesia is the world's third biggest producer of cacao (exceeded only by the Ivory Coast and Ghana). Approximately 1.5 million hectares (3.7 million acres) of the island nation are covered by cacao crops.

FAR LEFT:

Cacao pod, Bali, Indonesia

The cacao pod contains the beans inside in a protective wet pulp. The pod is opened by striking it with a wooden club, causing it to split apart along the centre.

LEFT TOP:

Cacao harvesting, Bali, Indonesia

Cacao beans develop inside large pods, which are harvested by cutting through the stalk either with a knife or (for pods higher up) with this elongated cutting sickle.

LEFT BOTTOM:

Cacao drying, Bali, Indonesia

Following a fermentation process, cacao beans are dried to take their liquid content down from approximately 60 per cent to 7.5 per cent. Here, dried beans are being sorted for quality.

OPPOSITE LEFT:
Latex tapping, Malaysia
Latex tapping from rubber trees demands skill and patience. The first stage is to cut spiral grooves into the bark of the tree trunk. Latex then oozes out of the cut for a period of about five hours.

OPPOSITE RIGHT:
Latex processing, East Java, Indonesia
Workers process latex at the state-owned rubber factory in Jember in Indonesia's East Java province, 2012. The milky fluid is refined, and can thereafter be manufactured into latex and other natural rubber products.

LEFT:
Rubber production, Thailand
The latex tapped from rubber trees is harvested into collecting cups. In these cups, the latex can be left to coagulate naturally upon exposure to air. This hardened product is then processed for sale.

RIGHT:
Rubber production, Thailand
Compacted blocks of coagulated latex are put through rollers of increasingly tighter compression to create more workable rubber sheets. This rolling action also squeezes out any excess moisture from the material.

OPPOSITE:
Rubber drying, Thailand
Rolled sheets of rubber are suspended from a wooden frame as the final stage of the drying process. The amber colour comes from accelerated drying over a smokey wood fire.

RIGHT:
Palm oil processing plant, Borneo, East Malaysia
Palm oil is a vegetable oil that derives from the fruit of oil palm trees. Palm oil has become a ubiquitous product, used in an estimated 50 per cent of the products we find in a typical supermarket.

OPPOSITE LEFT TOP:
Oil palm harvesting, Sabah, Malaysia
Oil palm kernels are harvested by a plantation worker and placed by the side of the plantation road, ready for collection and further processing elsewhere.

OPPOSITE LEFT BOTTOM:
Oil palm fruit
A close-up picture of ripe oil palm fruits. Globally, such fruits produce a greater volume of vegetable oil compared to any other type of vegetable oil plant.

OPPOSITE RIGHT:
Oil palm trees
A crop of oil palm trees, seen from above. The creation of oil palm plantations has been at enormous cost to tropical ecosystems, which have been cleared to make way for the profitable crop.

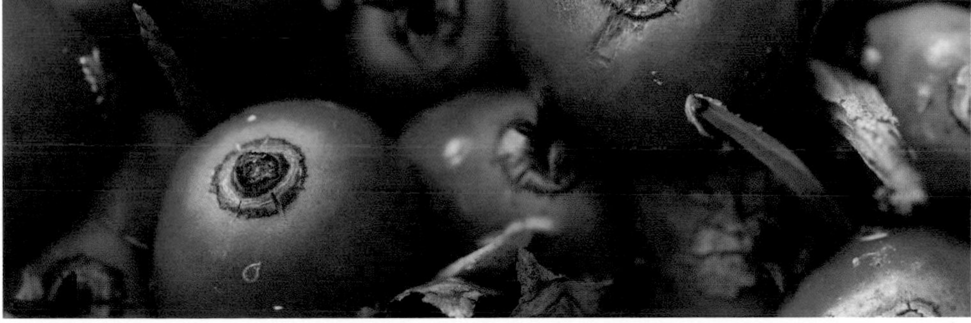

FAR LEFT:

Coconut trees, Malaysia

Coconut palm trees take about 15–20 years of growth to reach their full coconut productivity. Thankfully, they are flexible enough to withstand tropical storms and hurricanes without toppling or snapping, so they are a durable crop.

LEFT TOP:

Coconut harvesting, Malaysia

A worker harvests coconuts from high up in a palm tree. A less perilous method of harvesting is to cut the coconuts down using a sickle on the end of a long pole, worked from the ground.

LEFT BOTTOM:

Coconut harvesting, Malaysia

Two oxen stand by to transport a load of coconuts. The coconut itself is a mature seed, which contains the highly nutritious white flesh and the coconut milk extracted from the grated pulp. (The clear liquid inside coconuts is coconut water.)

LEFT:

Coconut harvesting, Thailand

Thailand is the world's sixth largest producer of coconuts, with major export markets globally, but especially to China – a voracious market that Thailand is struggling to adequately supply.

OPPOSITE:

Orchid farm, Thailand

The split dried coconut husks seen here will be used as natural 'pots' for the growing of orchid flowers. The husks are idea for orchid growing because that offer good water absorption and optimal root aeration, hydrating the orchid roots but avoiding making them soggy. The husks are also resistant to fungal and bacterial growth.

OPPOSITE:

Raisin production, Afghanistan

A farmer spreads grapes over the ground for air-drying into raisins. Depending on the size of the grapes, the dehydration process will take between one and three days.

BELOW:

Corn processing, Hanoi, Vietnam

A Vietnamese family works corn kernels off cobs with a basic powered threshing machine. Each ear of corn has roughly 800 kernels, and these constitute the main (although not the only) edible part of the plant.

RIGHT:

Corn drying, Vietnam

Dried corn is an ideal food product for the tropics, having a long shelf life even without refrigerated storage. Just 15 minutes of immersion in water is usually enough to rehydrate the kernels to an edible state.

RIGHT:

Sugarcane field, Quang Dien district, Vietnam

Sugar cane production thrives on an insatiable global market for sweetening agents, biofuels (ethanol) and bioplastics. The crop's intensive demand for water, however, has resulted in major environmental impacts, especially in Vietnam.

OPPOSITE:

Sugarcane harvesting, northern Queensland, Australia

An automated sugarcane harvesting process fills the air with dirt and debris. Workers in such conditions often experience respiratory tract and ocular infections. Current research also appears to indicate that sugarcane workers have a higher incidence of lung cancer.

RIGHT:
Grain storage, Victoria, Australia
Against the backdrop of an Australian sunset, these grain silos appear almost as works of modernist architecture. Silos are designed to keep the large volumes of grain inside cool and dry, which helps to prevent the propagation of fungus and bacteria.

OPPOSITE:
Grain harvest, Western Australia
The view from the cab of a combine working a large wheat field in western Australia. The name 'combine' (short for 'combine harvester') comes from the way that the machine combines three functions – reaping, threshing and winnowing.

LEFT:

Sheep farming, New South Wales, Australia

Sheep farming is an arduous business in Australia, most of the activity taking place over huge acreages of arid and semi-arid landscape. The most numerous of the sheep breeds farmed in Australia is the Merino, famed for the exceptional quality of its wool.

ABOVE:

Cattle mustering, Australia

These cattle have been mustered in pens, the neatness of the image hiding the hours of labour expended by stockmen, horses and cattle dogs to get them there. The world's biggest cattle station – Anna Creek in Australia – covers 24,000 sq km (9,266 sq miles).

ABOVE:
Stock farming, New Zealand
Agriculture is New Zealand's largest tradable economic sector, and over 6 million cattle are found across the nation. The cattle are mostly in dairy herds, and New Zealand is also the world's largest exporter of butter and whole milk powder.

RIGHT:
Sheep farming, Te Anau, New Zealand
New Zealand has a global reputation for its sheep farming, although its c. 28 million sheep today is a steep fall from the c. 70 million in the early 1980s. This drop is balanced out by the rise in the dairy cattle industry.

Africa and the Middle East

There are many challenges to farming in Africa and the Middle East, but by far the greatest of them is finding adequate supplies of water, and getting that water to crops and livestock. Africa, for example, has been an arid and drought-prone continent in many parts for centuries, but in recent decades an increasingly hot world has dried the land further, placing a greater strain on the region and its peoples. This is a critical threat in a continent where 65 per cent of the population are subsistence farmers. Only 7 per cent of Africa's arable land is irrigated properly, thus the remaining 93 per cent relies on annual rainfall to water crops and hydrate animals, and these rainfalls appear to be in decline, in the process lowering the levels of rivers and lakes.

The life of African and Middle Eastern subsistence farmers is usually invariably hard, almost indistinguishable from that of centuries or even millennia ago. In some countries, only 2 per cent of the population of rural areas have access to electricity, meaning that farming is done by the power of arms, legs and animals, with traditional tools. Yet Africa is also home to huge agricultural enterprises. Thus, despite environmental pressures, and uncertain futures, the farmers of Africa and the Middle East continue to provide themselves and the world with some bountiful crops, particularly ones that are much desired in the developed world – the region includes some of the biggest global producers of tea, coffee, fruits and cacao, which are shipped to markets across the globe.

OPPOSITE:
Cassava harvesting, Nigeria
After rice and maize, cassava (manioc) is the third most important source of carbohydrates in the tropical world. One of its key advantages is that it is highly resistant to dehydration, hence it is considered a 'food security crop' in regions regularly affected by extended droughts.

RIGHT:

Cassava plant, Nigeria
The cassava plant is able to grow in the most arid of regions, as we can see from this photograph. Rewatering after a long period of dry weather, however, will quickly revitalize plant growth.

FAR RIGHT:

Cassava tubers, Nigeria
The edible part of the cassava plant is its tuber, although this does require cooking to remove toxic or unpleasant elements. Like potatoes, cassava can be boiled, roasted, mashed or baked, and it can also be used in sweet desserts.

OPPOSITE:

Cassava harvesting, Nigeria
Nigeria is the world's biggest producer of cassava. By the beginning of the new millennium, it already acounted for 19 per cent of the entire global crop, rising to 21 per cent by the beginning of the 2020s.

OPPOSITE:

Wheat harvest, Al Minufiyah, Egypt

Egyptian farmers harvesting wheat in 2022. Egypt is the world's fifth largest wheat importer. However, since the Russia–Ukraine War has severely limited imports from Ukraine, Egypt has attempted to upscale its domestic wheat production in response.

LEFT:

Wheat winnowing, Tigray, Ethiopia

Ethiopian farmers winnow their crop of wheat, separating the wheat from the chaff by tossing it through an air stream. Like Egypt (see above), Ethiopia is planning a huge acceleration in wheat production in response to the Russia–Ukraine War and prolonged drought.

OPPOSITE:

Rice irrigation, Kirundo Province, north-east Burundi

A woman and her child stand next to a Chinese-sponsored irrigation system, pumping out the thousands of gallons of water required to flood the surrounding fields for rice growing. Inward investment from foreign countries has been crucial to many areas of African agriculture.

LEFT:

Rice seedlings, Mto Wa Mbu, Tanzania

Healthy rice seedlings are seen here growing in a rice nursery in Tanzania. This particular system is known as a wet-bed nursery, and it is planted with pre-germinated seeds. The seedlings are transplanted to a rice field after 15 to 21 days.

LEFT:

Rice fields, Nile River, Egypt

The Nile River has been a crucial source of irrigation in north-east Africa for millennia, and today 95 per cent of Egyptians live along the course of the river. Ominously, however, low rainfall levels are reducing the Nile's water levels year on year.

ABOVE:

Rice growing, Al-Beheira, Egypt

A farmer uses a horse to flatten the soil in preparation for growing rice in the Nile Delta province of Al-Beheira, north-west of Cairo. Because of ageing irrigation systems, the farmers often have to tap water from polluted canals.

OPPOSITE LEFT:
Sorghum field, Sudan
The African continent produces roughly 20 million tonnes (22 million US tons) of sorghum each year, and for many African people in food-insecure regions, sorghum is a critical crop for their survival.

OPPOSITE RIGHT:
Harvested sorghum, Uganda
Havested sorghum is left to dry out after the harvest in Uganda. Sorghum is a drought-resistant crop, thus it can grow successfully in many of the most arid regions of Africa.

LEFT:
Sorghum harvesting, Burkina Faso
Much African agriculture still depends upon human labour rather than expensive mechanization. These Africa women might have to transport the baskets of sorghum for many miles to their homes or market.

RIGHT:

Farming, Sudan

This Dinka farmer in South Sudan is tending his field of maize and other crops. The subsistence growing of grains (wheat, corn, rice) sometimes takes place alongside the production of cash crops such as sugar cane and cotton.

OPPOSITE:

Grain winnowing, Oromia district, Ethiopia

These Ethiopian farmers are winnowing teff, a drought-resistant edible grass seed that is a staple food in Ethiopia and neighbouring Eritrea. Teff is capable of delivering crop yields well in excess of wheat and many other grains.

RIGHT:

Maize farming, Witbank, South Africa

More than 15 million tonnes (16.5 million US tons) of maize are grown in South Africa every year, the output growing hugely in recent years to fuel national and global demand. Here, the ripe maize is ready for harvesting, a combine harvester waiting in the background.

OPPOSITE:

Maize field, South Africa

Maize grows at its best in sunny, warm climates, hence South Africa offers excellent growing conditions for the crop, rain depending. About half of Africa's maize crop is used in animal feed and half for human consumption.

OPPOSITE:
Cacao beans, Ghana
Cacao is the dominant cash crop of Ghana, the second largest producer of cacao in the world. An unfortunate side effect of this industry is heavy deforestation, with large tracts of primary rain forest being cleared to make way for cacao growing.

LEFT:
Drying cacao beans, Ivory Coast
Cacao beans are spread out to dry under the African sun. Mechanization is still low in much of Africa's cacao industry, and it is estimated that some 2.1 million children are used as labour in West African cacao farming.

RIGHT:

Drying cacao beans, Ghana
A Ghanaian farmer spreads out cacao beans to dry. The drying process takes several days, and follows a fermentation stage. Once the beans are dried, they can be packed and taken to market.

OPPOSITE LEFT:

Unloading cacao, Rouen, Normandy, France
The international trade in cacao is a vast enterprise, with the global cacao and chocolate market valued at around USD 46.61 billion (c. £40.5 billion) in 2021. Here, a mechanized grabber is used to offload a shipping container full of cacao beans in Rouen, France.

OPPOSITE RIGHT:

Cacao exports, Ghana
Ghanaian dock workers load up sacks of cacao beans ready for export. The huge growth in the cacao market is primarily driven by the insatiable global demand for chocolate confectionary.

BENOTO

Fry
GHANA COCOA

RIGHT:

Drying coffee beans, Kiambu, Kenya

On a Kenyan coffee plantation, coffee beans are laid out to dry in the sun, supported in wooden drying beds. About 95 per cent of Kenya's coffee is sold into export markets.

OPPOSITE LEFT:

Coffee beans, Kenya

Ripe coffee beans hang thick on this Kenyan coffee tree. A major plantation will usually have coffee trees at a density of 1,200–1,800 plants per hectare (500–750 plants per acre).

OPPOSITE RIGHT:

Coffee plantation, Kenya

On a coffee plantation outside Nairobi, women carry containers brimming with coffee beans to the factory for hulling (removing the coffee beans from their fruit) and subsequent processing.

KEPCHO

Tea farming, Africa
Tea growing is a major agricultural enterprise for many African countries, including Malawi, Uganda, Tanzania, Rwanda, Mozambique and Burundi. However, Kenya stands clear as the largest volume producer, harvesting more than half a million metric tons of tea every year.

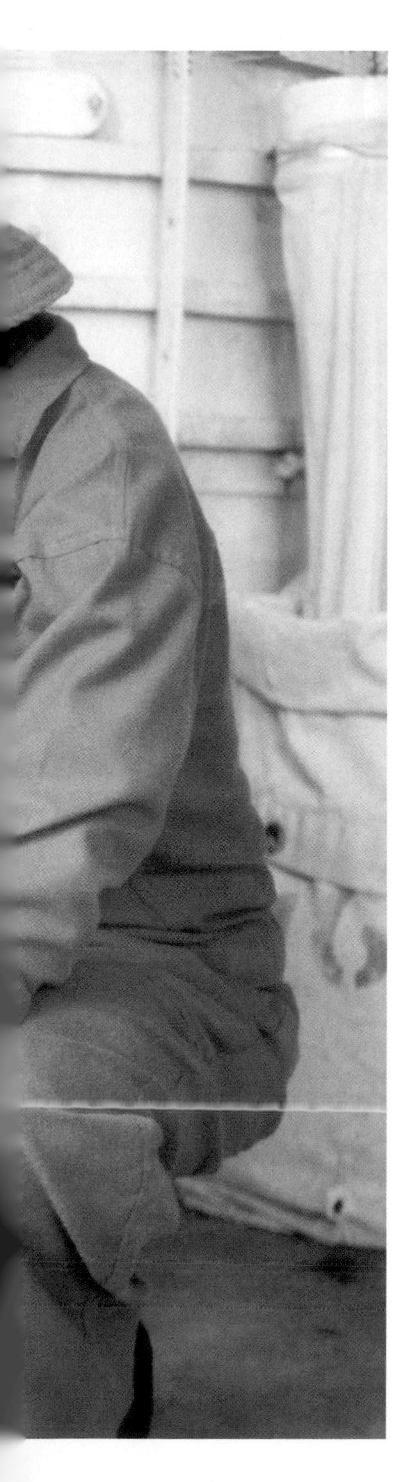

OPPOSITE:
Tea sacks, Kericho, Kenya
The town of Kericho is a centre for the Kenyan tea trade, being the focal point for the Kericho Tea Plantation, the largest tea plantation in Africa. The estate covers 3,339 hectares (8,250 acres) and employees 16,000 workers.

FAR LEFT:
Tea inspection, Kericho, Kenya
A Kericho plantation worker inspects black tea as it comes from drying.

LEFT:
Grading tea, Mombasa Kenya
The activity of tea grading involves classifying the tea according to leaf size and type, with some further criteria relating to the shape, condition and completeness of the leaf.

BWR
371MP

LEFT:
Orange harvesting, South Africa
These workers have harvested a healthy crop of oranges. An orange tree, aged between 10 and 40 years old, can produce anywhere between 250 and 1,000 fruits every year, depending on the climate, pest control and the experience of the orange farmer.

ABOVE:
Orange trees, South Africa
A tractor pulls a pesticide spraying unit down rows of orange trees. There are numerous pests and pathogens that can spoil an otherwise robust orange crop. The viral disease *Citrus tristeza*, for example, can kill a healthy tree in three months.

FAR LEFT:

Mechanized grape harvesting, South Africa

Grape harvester vehicles are the ultimate tools for harvesting large commercial vineyards. With their high ground clearance, the vehicles can straddle a row of fully grown vines, performing vine shaking and collection in a single pass.

LEFT:

Manual grape harvesting, South Africa

Manpower is a more common method of grape harvesting across Africa. The ripe bunches of grapes are cut with shears and placed in harvesting crates (like the one seen here), the contents of which are deposited in large tractor-pulled bins with a capacity of several tons.

RIGHT:

Wine making, Stellenbosch, South Africa

Large volumes of grapes are turned through a crushing machine, which crushes the grapes to separate the juice from the flesh and to destem the fruit. This dual process can alternatively be performed via a pressing machine, or sometimes crushing and pressing are performed sequentially.

FAR RIGHT:

Wine fermentation vats, South Africa

In wine making, fermentation is both a natural and an artificially induced process, one that eventually turns the sugars of the fruit into alcohol. The process takes some time, typically about 10 days to a month. In commercial wine production, it takes place in scientifically controlled conditions.

RIGHT:
Sheep shearing, South Africa
Sheep are sheared not only for their wool (a product in itself), but also to keep the sheep clean, comfortable and with reduced incidences of diseases and parasite infestations. This farmer is shearing a large sheep with a traditional pair of hand-operated sheep shears.

OPPOSITE:
Sheep farming, Swartland, South Africa
Sheep farming is big business in South Africa, with approximately 8,000 commercial sheep farms throughout the country, farming nearly 30 million sheep. Here, two young workers help herd a small flock into a pen.

INTERNATIONAL

RIGHT:
Camel caravan, Sahara Desert, Africa
Camel farming is a growth industry in Africa, the animals providing a nutritious and low-fat milk that is becoming increasingly popular in global markets. Camels also provide meat.

OPPOSITE:
Camel farming, Chad
Camels take a drink from a water trough on a hot day in Chad. Because camels can cope with drought conditions for prolonged periods, many African farmers are switching from cattle farming to camel farming. A camel can endure a three- or four-day trek between wells without adverse health effects. Camels can also forage for rough grasses and plants.

OPPOSITE:

Village well, Mali

These farmers are using a donkey to operate a traditional well, raising water that will be used to quench the thirst of the surrounding sheep, goats and camels, as well as the humans. Africa's volume of groundwater far exceeds that of its above-ground freshwater.

LEFT:

Goat milking, Mauritania

An African goat herder milks one of his flock by hand. Goats are ideal creatures for subsistence farmers or small-farm systems, being hardy and easily fed and cared for, while producing milk, meat and skins.

RIGHT:
Cattle farming, Niger
Throughout African history, cattle farming has been about far more than simply managing sustainable food herds. Cattle have been a form of tribal and imperial currency, the volumes of cattle owned by a people often equating to their political power and territorial extent.

OPPOSITE LEFT:
Cattle herder, South Sudan
Some of the breeds of Sanga cattle of South Sudan are known for their incredible horn profiles. The horns can grow up to 2.4 m (8 ft) in length, giving the creatures a majestic appearance.

OPPOSITE RIGHT:
Cow herd, Chad
Finding adequate water supplies is often one of the greatest challenges of cattle farming in Africa, especially in the dry season. The problem is exacerbated by the fact that the livestock are also in competition with crop irrigation over scarce water supplies.

OPPOSITE:
Goat herding, Somalia
About 80 per cent of the population of Somalia are pastoralists, meaning that they manage herds of domesticated animals in grassland environments, the herders typically moving with their creatures in nomadic fashion between feeding and watering areas.

LEFT:
Sheep farming, Kenya
The scarcity of water that threatens pastoral farmers is evident in this photograph. In 2009, when this photograph was taken, Kenyan nomadic herders near El Wak walked up to 45 km (28 miles) with their herds every few days in search of water, after seasonal rains failed.

North America

When we think of North American agriculture, we tend to think big. The USA in particular has seemingly endless tracts of agricultural land, from grain-growing prairie farms that stretch from horizon to horizon to luxurious vineyards that soften acres of Californian hillsides and valleys. At the same time as the USA has the land, it also has the money and the machines to farm it. Indeed, the USA displays the most advanced large-scale agricultural mechanization and automation to be found anywhere in the world, meaning that the country has a lower agricultural worker-to-acre ratio that any other nation. In total, the USA has more than 2 million farms across 3,626,000 sq km (1,400,000 sq miles) of farmland. Across that land, and generalizing heavily, the major crops are maize, wheat and soybean within the Great Plain regions, fruits, vegetables and nuts in California and Florida, and cotton, tobacco and rice in the south. Livestock farming crosses the country, with heavy concentrations in the west and south-east.

Canada's agricultural situation is quite different to that of the USA. Its harsher northern climate and tough soils mean than only 7 per cent of its land is suitable for crop agriculture, although given the size of the country, this is still an impressive swathe of territory for crops and livestock. Both Canada and the USA are negotiating climate change, however, particularly the USA regarding water supplies, alongside wildfires and extreme weather events, as elsewhere in the world.

OPPOSITE:
Soybean harvesting, Paris, Illinois, USA
A combine dumps beans into a grain wagon as farmers harvest a soybean field east of Paris, Illinois. This aerial view perfectly demonstrates the synergy between tractor and combine during mechanized harvesting. Note the crop residue being blown out of the back of the machine by the straw chopper.

Prairie farm, Alberta, Canada
The province of Alberta in Canada has vast, rolling areas of farmland, with a total farm area of 20 million hectares (49 million acres) in 2021. Just over half of this area consists of land given over to hay and field crops, typical of this picturesque prairie scene here.

LEFT:

Grain silos, Alberta, Canada
These tower grain silos protect the crop inside from the cold and wet outside. They are made from galvanized steel, which has many advantages, such as low weight, easy construction and minimal maintenance, but can cause condensation build-up inside if the silos aren't monitored and controlled.

OPPOSITE:

Grain elevator, Hudson Bay, Saskatchewan, Canada
A grain elevator is not merely a place in which grain is stored, but also one from where grain can be moved on to transport for distribution.

SASKATCHEWAN
HUDSON BAY
A

ABOVE:
Wheat, Alberta, Canada
Across our planet, more agricultural land is devoted to wheat growing than any other crop, with a global wheat production of 761 million tonnes (839 million US tons) in 2020. Canada's wheat output is forecast to leap by more than 50 per cent.

RIGHT:
Wheat harvesting, North Dakota, USA
These John Deere combine harvesters provide the ultimate vision of mechanized agriculture. Each machine is capable of harvesting up to 7,200 bushels of corn each hour, which is enough to fill seven trucks.

LEFT:

Wheat seed, Saskatchewan, Canada

Wheat seed is moved from a semi-trailer into an air drill for seeding. Air-drill seeders are machines designed both to till and seed the ground in a single pass, saving the farmer time by combining separate processes. This is invaluable on farms with very extensive acreages.

ABOVE:

Grain harvest, Saskatchewan, Canada

Large industrial farmers have access to highly advanced technologies for monitoring the condition of stored grain, including sensors for measuring moisture, air temperature, grain temperature and CO_2 production, the data being fed wirelessly to a computerized monitoring station.

Prairie farm, Saskatchewan, Canada
This lofty photograph evokes the sheer scale of North American prairie farms. Individual farms can cover tens of thousands of acres. This one is given over to wheat fields. About 34 per cent of Canadian agricultural land is devoted to grains and oilseeds, followed by some 24 per cent to livestock.

ALL PHOTOGRAPHS:
Canola crop, Ontario, Canada
The seeds of the canola plant produce an oil used in biofuels and foods such as margarine and cooking oil. In the large photo here, a canola field is watered by a huge central pivot irrigation system – essentially a large, automated sprinkler system that turns around a central anchor point.

VALLEY
8000 series

RIGHT:
Planting corn seed, Illiopolis, Illinois, USA
Each of the yellow hopper bins here contain wheat seed. The wheat seeder machine automatically ploughs the ground and plants the seed to a required depth and at required intervals.

OPPOSITE:
Wheat harvesting, Indiana, USA
Harvested wheat pours into a grain truck working alongside a combine. Indiana has exceptional quality of soils for grain crops, and corn and soybeans are the two larger crop types in the state. The specific type of wheat grown in Indiana is known as soft red winter wheat.

22

OPPOSITE:

Crop dusting, Montana, USA

The sheer scale of many of North America's prairie farms means that aerial crop dusting is often the most cost-effective and fastest method of applying pesticides to crops. Crop dusting is a highly skilled feat of flying, therefore most farmers contract out the service to specialized aviators.

LEFT:

Sunflower field, North Dakota, USA

The USA produces about 1 million metric tons of sunflowers each year, although even this crop size places it only 10th in global sunflower production; Russia and Ukraine hold first and second positions respectively.

OPPOSITE:
Farmland, Kansas, USA
Kansas is the quintessential, and largest, agricultural state in the USA. This aerial view of winter fields clearly shows the strange circular patterns of crop growth that are generated by centre pivot irrigation technologies.

LEFT:
Grain silos, Kansas, USA
Huge grain silos dot the landscape throughout Kansas, which is the largest wheat-farming state in the USA. This single state accounts for about 18 per cent of all US wheat production and a massive 55 per cent of all grain sorghum production.

RIGHT:

Ploughing, Minnesota, USA

For every mega-farm in the USA, there are dozens of far smaller operations, often family-run. Here, a lone farmer uses tractor and plough to turn over a field of corn stubble, loosening compacted soil to enable further growing.

OPPOSITE:

Draught horses, Amish Country, Ohio, USA

The Amish communities of the USA shun modernity in their way of life, extending that principle to their agricultural practices. These five sturdy draught horses, yoked to a plough, are working animals.

LEFT:

Cowboy, Alberta, Canada

The cowboy has long been an iconic figure in North American history, synonymous with a hard, free life and the settlement of the American West. Here, a Canadian cowboy watches over his cattle herd.

ABOVE:

Roping, Wyoming, USA

Although roping has become both entertainment and sport, for cowboys it is a hard-won working skill they use frequently. Because open ranges have no corrals, roping is often the only way for a cowboy to work on an individual cow or horse.

RIGHT:
Ranch, Western Colorado, USA
Many US ranches sit in areas of breathtaking natural beauty, albeit places that are often unforgiving to both humans and animals. A rancher will labour in these conditions from before dawn until well after dusk, seven days a week.

OPPOSITE:
Bull, Manitoulin Island, Ontario, Canada
The bull here is of the French Charolais breed, a type of high-yielding beef cattle that has spread well beyond the borders of France, being farmed today in 68 countries internationally, including the USA and Canada.

OPPOSITE TOP:

Cotton plants, Georgia, USA

Cotton is a natural plant fibre that has been spun and woven into clothing and fabric products since at least 3000 BCE. About 31 per cent of the world's textile market belongs to cotton.

OPPOSITE MIDDLE:

Monitoring cotton plants, Georgia, USA

Each cotton boll contains approximately 27–45 seeds, and each seed has some 10,000–20,000 tiny fibres attached, each only about 1 cm (½ in) in length.

OPPOSITE BOTTOM:

Cotton crop, Fort White, Florida

In modern cotton farming, the harvested cotton is rolled into a tight bale, protected by an outer layer of plastic wrapping to protect the cotton from the weather.

LEFT:

Cotton harvesting, Georgia, USA

Cotton-picking machines are separated into stripper and picker types, the former stripping off almost the entire plant, the latter machine removing only the cotton from the open bolls.

Coke
FARMS
PARKHILL

OPPOSITE:

Tobacco farming, South Carolina, USA

Farmers harvest a tobacco crop on a family-owned farm in South Carolina. Despite the now well-documented health concerns associated with tobacco, it remains a profitable crop option for both small- and large-scale farmers, at least those who live in the optimal climate band for growing the crop.

FAR LEFT:

Tobacco harvesting, Kentucky, USA

Brightleaf tobacco leaves are here tied to stakes at harvest prior to being transported to the curing barns. Tobacco is the most lucrative crop per acre in all agriculture. Whereas an acre of wheat might sell for $300, the same area of tobacco would sell for around $1,500.

LEFT:

Tobacco drying, North Carolina, USA

An old tobacco-drying barn is used to air cure burley tobacco in Prices Creek, North Carolina. The three main tobacco types are Virginia, burley and oriental, each of which has its own smoking characteristics.

OPPOSITE LEFT:
Aquaponics system, USA
Aquaponics is a growth system that accounts for a widening diversity of crops in the USA. It is a very water-efficient system, which is why it appeals to farmers working in drought-hit states.

OPPOSITE RIGHT:
Radish plant grown using aquaponics, USA
Aquaponics is a system of agriculture that combines hydroponics and aquaculture to raise plants and fish in a closed system: the waste produced by the fish provides nutrients for the hydroponically grown plants, and the plants clean the water for the fish. This radish has been grown in this way in an aquaponics tank.

RIGHT:
Catfish farming, Mississippi, USA
Catfish farming is a major form of aquaculture in the USA. The three states in which the industry is concentrated are Mississippi, Arkansas and Alabama, with Mississipi the leader in a multi-billion dollar industry.

Vineyard, Sonoma County, California, USA
California is one of the world's largest wine-growing territories. A specific US wine region is often classified as an American Viticultural Area (AVA), meaning that 85 per cent of the grapes listed on the bottle are grown within the AVA. Sonoma County has no fewer than 13 AVAs, and is the most productive of the various California wine-growing regions.

RIGHT:
Peanut crop, Georgia, USA
A farmer inspects his peanut crop in Georgia. Peanuts can be a sensitive crop, best grown in warm and moist conditions, avoiding extremes of temperature.

OPPOSITE LEFT TOP:
Peanuts, Georgia, USA
Peanuts are classified as both a grain legume and an oil crop. Each peanut pod (the fruit of the plant) contains one to four edible seeds.

OPPOSITE LEFT BOTTOM:
Pistachio nuts, California, USA
Pistachio nuts mature on the tree in San Joaquin County, Sacramento Valley, California, the state that accounts for 98 per cent of all the pistachios grown in the USA.

OPPOSITE RIGHT:
Almond trees, California, USA
An astonishing 80 per cent of the world's almonds are grown in California, but prolonged drought conditions are presenting a severe long-term existential threat to this profitable industry.

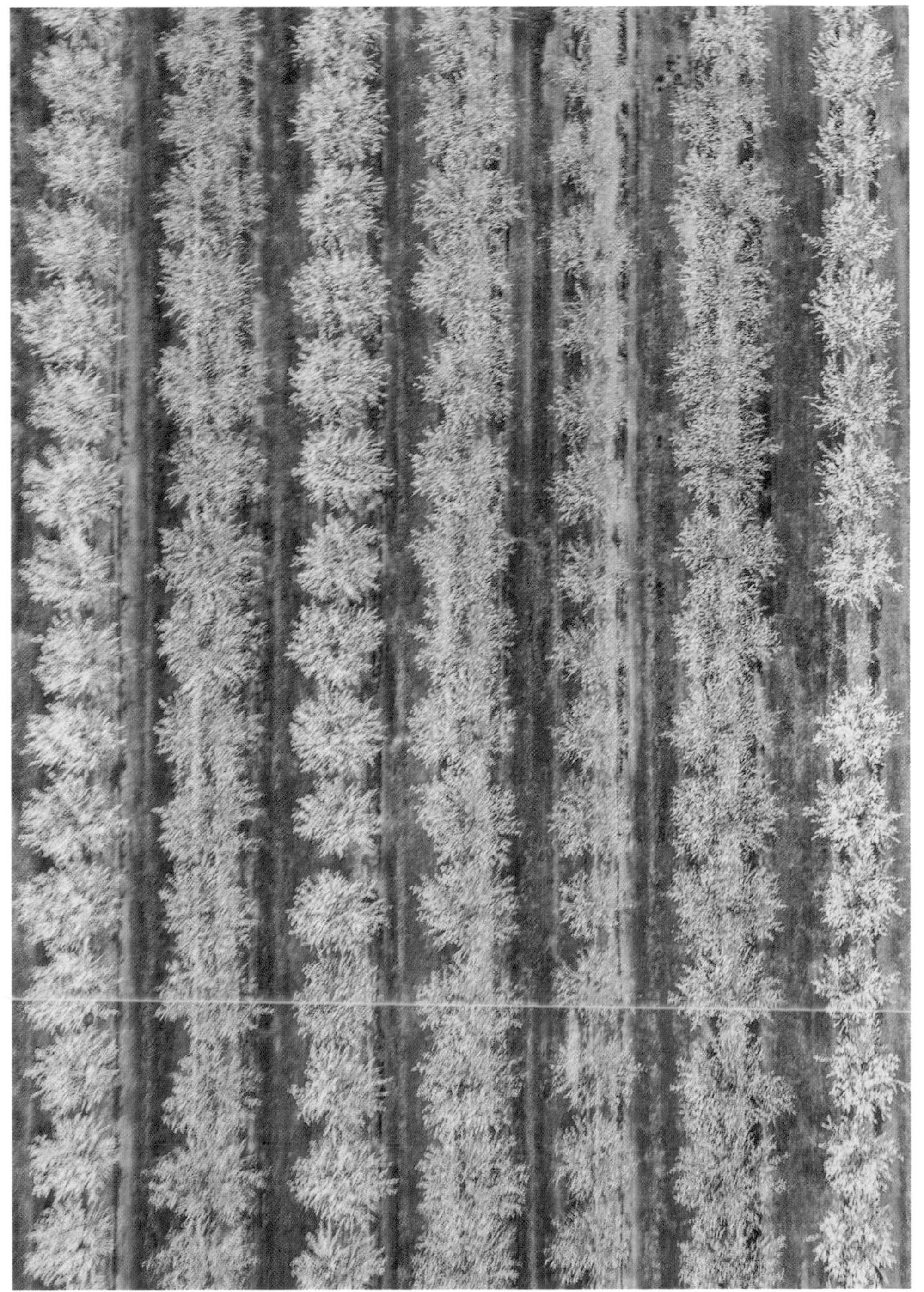

OPPOSITE:

Olive plantation, California, USA

Olive farming is a business that requires much patience. It can take three to five years for a new olive tree to start bearing fruit and some trees might only produce significant crops every sixth or seventh season.

LEFT:

Oranges, California, USA

Oranges grow readily in California's predictably sun-rich environment. Indeed, such was the agricultural explosion in orange fruit growing in the 19th century that it has been referred to historically as California's 'second gold rush'.

CHLORINE WATER
AGUA CON CLORO

OPPOSITE:
Lettuce harvesting, Imperial Valley, California, USA
Agricultural workers harvest a crop of Californian iceberg lettuce, placing the lettuces on to the belt of a mobile processing line.

LEFT TOP:
Cranberry bog, Cape Cod, USA
Cranberries grow in wetland bogs, the fruits clinging to vines that extend horizontal stems, called runners, through the wet, peaty ground. The growing season extends from April through to November.

RIGHT TOP:
Cranberry harvesting, USA
Bright red cranberries float in a flooded bog during the annual autumn cranberry harvest. The wet harvesting process involves flooding the cranberry bog then dislodging the berries from the plants so they rise to the surface.

LEFT BOTTOM:
Cranberry harvest, Massachusetts, USA
During the wet harvesting of cranberries, the farmers wade through the bog and gather the fruit using large wooden or plastic brooms, containing the collected fruit in a process called corralling

RIGHT:

Pumpkin patch, California, USA

Pumpkin growing is found across the USA, but the bulk of the industry is concentrated in California, Illinois, Indiana, Michigan, Texas and Virginia. The most productive farms can grow approximately 18,200 kg (40,000 lb) of pumpkins per acre of land.

FAR RIGHT:

Irrigation, California, USA

California is facing an acute water shortage, with a political and environmental battle being fought over whether so much water should be devoted to agricultural production – big agriculture accounts for about 80 per cent of all water use in the state.

Central and South America

Historically, Central and South America are regions of great fertility, abundant natural growth fuelled by tropical sunshine, rich soils, predictable rainfall (in parts) and a population skilled in farming. Together, they constitute one of the largest agricultural exporters in the world, aided by the fact that the richest nation on earth, the USA, is a proximate destination for its crops and meat, including coffee, cacao (although from 2020 this crop was crippled by a devastating fungus), cashew and Brazil nuts, pineapples, bananas and beef.

There are many question marks regarding the environmental cost of Central and South America's endlessly swelling agricultural production. Since the mid-1980s, to take just one example, global banana demand has swelled by more than 175 per cent, and deforestation of Latin American virgin tropical jungle has been consequent on meeting this demand. Around 1.9 million hectares (4.8 million acres) of the Amazon rainforest were lost in 2020 alone, and agriculture is one of the primary causes, particularly cattle ranching and small-scale subsistence farming. The biodiversity of Central and South America is ultimately the region's greatest natural asset, and must be preserved not just for the maintenance of irreplaceable natural expanses, but also for the viability of future agricultural production. The fact remains that Central and South America provide much of the world with the tropical crops we so enjoy, so it is in all our interests to see this remarkable farming region thrive.

OPPOSITE:

Coffee plantation, Vila Mariana, São Paulo, Brazil

A plantation worker captures ripe coffee beans in a net in the orchard of the Biological Institute, an applied research centre and the oldest urban coffee plantation in Brazil, located in Vila Mariana, south of São Paulo. The Institute's activities have included researching ways of tackling coffee plant pests.

RIGHT:

Coffee plantation, Hacienda San Alberto, Colombia

A place of exquisite beauty, the Hacienda San Alberto is also one of Brazil's premier coffee plantations, its coffees winning more awards than any other plantation in Colombia. The plantation's coffees can be tasted in any of several 'Coffee Temple' coffee shops dotted throughout urban Colombia.

OPPOSITE:

Coffee harvesting, Bahia State, Brazil

Coffee production in Brazil has experienced significant growth in recent years, from 48 million 1 kg (2.2 lb) bags in 2010 peaking at 65 million bags in 2020. Brazil is the world's largest coffee producer, but the international market is now much more competitive.

LEV
ORTLEV

OPPOSITE:

Coffee washing, Bahia, Brazil

A crop of coffee beans is washed following harvesting. A 'washed' or 'wet' coffee is one that has had various fruit layers removed before drying begins, giving a more consistent flavour. This contrasts with 'dry' or 'naturals' coffees, which are dried in the full cherry prior to de-pulping. These give more fruity and fermented flavours, but there can be variable and sometimes unpleasant tinges.

LEFT:

Tobacco plantation, Viñales Valley, Cuba

Cuba is regarded as producing the world's finest tobacco, and especially for the manufacture of prestigious brands of cigar. Every year, some 100 million cigars are made in Cuba, many hand-crafted.

OPPOSITE:
Cacao harvest, Cuernavaca, Colombia
Coffee farmers separate pulpy cacao seeds from cacao pods during a harvest in 2021. Although in English we use the word 'cocoa' to describe both bean and plant, technically 'cacao' is the unprocessed bean and 'cocoa' is the processed product.

LEFT:
Organic cacao beans, Mindo, Ecuador
Unroasted cacao beans are now often used in the production of vegan chocolate, as they have been through minimal processing before the conversion into confectionary. Here, cacao beans are being dried following harvesting.

LEFT:

Sugarcane plant, Brazil

Sugarcane is a thirsty crop, requiring 2,000–2,300 mm (80–90 in) of water during the growing season, although the soils need to be well drained. Every 100 tonnes (110 US tons) of cane produces about 12 tonnes (13 US tons) of sugar and 4 tonnes (4.4 US tons) of molasses.

RIGHT:

Sugarcane burning, Bariri, Brazil

Sugarcane fields are often burned prior to harvest to destroy the large volumes of dried leaves (known as 'leaf trash') that builds up and which can impair the growth of the next season's crop. Burning also makes harvesting a cheaper process. The smoke emissions, however, are known to cause respiratory ailments.

OPPOSITE:
Soybean pods, Brazil
The mature tan colour of these soybean pods means that they are ready for harvesting. The ideal moisture content level for the plants at harvest time is 13–15 per cent; any lower than that and they risk splitting or shattering under the harvesting process.

LEFT:
Soybean harvesting, Brazil
Soybean production has seen stunning growth in Brazil, the country overtaking the USA as the world's biggest producer in the 2017–18 season. In the 10 years between 2010 and 2020, the Brazilian soybean acreage expanded by 60 per cent.

ABOVE:
Yoke of oxen, Andes, Venezuela
This Venezuelan mountain farmer is using oxen to plough a potato field. These huge, powerful creatures have excellent traction and stability, even in highly sloping areas of farmland.

OPPOSITE LEFT:
Potato planting, Andes, Venezuela
Potato farming is a tiny enterprise in Venezuela. Furthermore, a lack of seeds and fertilizers resulted in a precipitous 70 per cent drop in the area of land planted with potatoes in 2018.

OPPOSITE RIGHT TOP:
Potato harvesting, Peru
Potato harvesting the hard way – kneeling on the ground and pulling up potatoes by hand. Increasing global temperatures have meant that Andean farmers have faced crop diseases that were previously usually confined to warmer regions.

OPPOSITE RIGHT BOTTOM:
Potatoes, Urubamba Valley, Peru
Potatoes are the main crop of more than 80 per cent of farmers working the land in the Peruvian Andes. Traditional potato types common in the Andes have become fashionable delicacies on international markets.

OPPOSITE:

Field of corn, Colca Canyon, Peru

A field of corn thrives in a high-altitude location in the Peruvian Andes. More than 50 different types of corn originated in Peru. One of the most popular types eaten in Central and South America is *choclo*, a large-kernel variety with a chewy, starchy texture, more savoury than sweetcorn.

LEFT:

Maize, Chilcapamba, Ecuador

An Ecuadorian farmer inspects his maize crop prior to harvest. Maize is a ubiquitous crop in Ecuador, although there have been recent governmental moves to push farmers towards growing other types of crop that are more profitable on the export market.

Cattle herding, Patagonia, Argentina
In South America, the *gaúcho* is as legendary as the cowboy of the North American West. The *gaúcho* is primarily a cattle-herder by trade, but his title evokes deep cultural qualities, such as bravery and nobility alongside a wily intelligence. Here, modern-day *gaúchos* herd cattle on a range in Patagonia.

ABOVE:
Cattle farm, Argentina
Argentina is known for its huge cattle industry, focusing mainly on the production of beef. The cattle farms range from small family-owned enterprises with a few dozen head of animals through to massive commercial ranches over thousands of acres.

OPPOSITE:
Llamas, Andes, Bolivia
Camelid farming (llamas and alpacas) is a crucial industry for tens of thousands of poor rural families in Bolivia, especially those concentrated in the Bolivian Altiplano (high plateau). These easily domesticated animals are farmed for their fleeces, meat and hides.

LEFT:
Pig farming, San Francisco El Alto, Guatemala
A woman pets her sow, swarming with a healthy litter of piglets. For subsistence farmers, such a litter can make a profound difference to the family table and to family finances, as some of the litter can be sold on.

OPPOSITE:
Vineyard, Mendoza Province, Argentina
Argentina is today the fifth largest wine producer in the world, with export sales alone (just 10 per cent of the total output) amounting to about US$ 900 million in value. Some 72 per cent of all the wine produced comes from the Mendoza region, in the eastern foothills of the Andes.

OPPOSITE:
Vineyard, Mendoza Province, Argentina
Women harvest a luxurious crop of grapes at a vineyard in Luján de Cuyo, a wine sub-region of Mendoza producing wine types such as Malbec, Cabernet Sauvignon, Chardonnay and Torrontés.

RIGHT:
Vineyard, Valle del Elqui, Chile
Many of the South American vineyards, including this one in the Valle del Elqui in Chile, grow at high altitudes in arid conditions, the climate characterized by intense sunlight during the day and chilly nights. The Valle del Elqui is known for its Syrah, Cabernet Sauvignon, Chardonnay and Pinot Noir.

OPPOSITE:

Blue agave harvesting, Tequila, Jalisco, Mexico

In a scene that would have looked little different a hundred years ago, two farmers harvest the blue agave plant, the core ingredient of the tequila drink, in turn the base of the internationally famous margarita drink. Tequila is also the name of the town in which the alcohol is produced.

LEFT:

Agave processing, Tequila, Jalisco, Mexico

In the manufacture of tequila, the large *piña* bulb is baked in an oven to extract its fermentable sugars. An alternative to the large stainless-steel ovens seen here are traditional clay-and-brick types called *hornos*.

OPPOSITE:

Banana plantation, Costa Rica

A conveyor system is used to transfer harvested ripening bananas through this large banana plantation in Costa Rica. Costa Rica exports bananas to more than 35 countries globally.

LEFT:

Banana processing, Costa Rica

Workers wash and prepare bananas for shipment. When bananas are being exported to foreign markets, shipping conditions must be cool and dry to prevent the bananas ripening and then rotting during transit.

ABOVE AND RIGHT:
Coconut plantations
Coconut plantations proliferate across Central America and the northern parts of South America, being mainly confined to a strip of the planet situated in tropical regions from 25 degrees north latitude to 25 degrees south latitude.

OPPOSITE:
Pineapple crop
Pineapple crops thrive best in areas of heavy and predictable rainfall, hence many humid tropical areas of Central and South America are ideal. The plant can also grow in almost any type of free-draining soil.

Picture Credits

Alamy: 11 left (Agencja), 12 (Jochen Tack), 13 (David Bagnall), 16 (Mark Richardson), 21 (Karen Appleyard), 22/23 (Realimage), 27 (Loop Images), 28 (MS Bretherton), 35 top left (Chritiano Gala), 35 bottom left (Marc Hill), 35 right (imageBROKER), 41 (Elena de las Heras), 42 (Martin Apps), 43 top left (Matthew Richardson/Stockimo), 43 bottom left (Evan Bowen-Jones), 43 right (David Wootton), 49 (Irina Naoumova), 54 (Robert Hoetink), 55 (Robert Harding), 56 (UJ Alexander), 57 right (Jeff Morgan 16), 58 (Alex Ramsay), 59 (Laurence Jones/ljonesphotography.co.uk), 60 left (Geo-grafika), 60 right (UschiDaschi), 68 (Horizon Unternational Images), 71 top (Universal Images Group North America), 71 bottom (Ashley Cooper), 73 left (Top Photo Corporation), 73 bottom right (Hemis), 77 (Reuters), 79 (Zoonar), 90 right (Reuters), 94 (Scubazoo), 97 bottom right (Malcolm Price), 102 (Vu Khoa Nguyen Khanh), 106 (Horizon International Images), 108 (Uwe Moser), 112 left (Amar & Isabelle Guillen-Guillen Photo), 115 (Sean Sprague), 116 (Xinhua), 117 (Charles O Cecil), 118 (Amar & Isabelle Guillen-Guillen Photo),119 (Reuters), 120 left (MJ Photography), 121 (Joerg Boethling), 128 (Robert Harding), 129 right (Superstock), 131 right (Marion Kaplan), 134 left (Peter Jordan), 134 right & 135 (Kirsty McLaren), 136 (Afripics. com), 139 (Frans Lemmens), 140 (Design Pics Inc), 142 (David South), 143 (Eckhard Supp), 146 (Greenshoots Communications), 148 (Frans Lemmens), 149 left (Media Drum World), 150 (The Color Archives), 151 (Reuters), 159 (GH Photos), 161 (All Canada Photos), 167 (Dennis MacDonald), 170 (Martin Shields), 171 (David Mabe), 172 (Grant Heilman Photography), 174 (Design Pics), 177 (All Canada Photos), 178 middle (Inga Spence), 178 bottom (Pat Canova), 180 left (Dennis MacDonald), 181 (Richard Ellis), 183 (Grant Heilman Photography), 186 & 187 top & bottom left (Inga Spence), 190 (Ellisphotos), 196 (Chester Voyage), 208 & 209 (Jesse Kraft), 210/211 (Y Levy), 213 (Olga Tarasyuk), 214 (David Litshel), 215 (Efrain Padro), 216 (Yadid Levy), 220 (Marko Reimann), 221 (Central America), 221 right (James Quine)

Dreamstime: 17 (Bengt Kohler Sandberg), 30 (Vaclav Volrab), 37 bottom left (ChenZhou), 37 right (Freeprod), 40 right (David Harraez), 45 top right (Jeanette Lambert), 47 (Evgeny Kharitonov), 52 (Kloeg008), 61 (Chris Moncrieff), 66 (Zzvet), 67 (Aliaksandr Mazurkevich), 86 (Sergii Figurnyi), 89 left (Saiko3p), 89 top right (Dodohawe), 90 left (Phunmead), 92 (Aleksandr Frolov), 93 (Platongkoh), 95 top left (Brian Scantlebury), 95 bottom left (Shuttersyndicate), 101 left (Alexander Podshivalov), 103 (John Krutop), 105 (Christopher Bellette), 120 right (Ava Peattie), 123 (Sjors737), 132/133 (Jensphoto999), 141 (Hel080808), 157 (Scott Prokop), 158 (Fallsview), 160 (Nancy Anderson), 182 both (JosefKubes), 184/185 (Galina Barskaya), 187 right (Durson Services Inc), 189 (Welcomia), 192 (Patricia Marraquin), 193 (Welcomia), 194, 197 & 198 (Alf Ribeiro), 199 (Sabino Prente), 202 & 203 (Alf Ribeiro), 206 & 207 left (Walter Otto)

Getty Images: 75 both & 101 (Javed Tanveer), 110 (Bloomberg), 112 right (AFP), 113 (AFP Contributor), 114 (Anadolu Agency), 122 (Eye Ubiquitous), 137 (Martin Harvey), 200 (Jan Sochor)

Shutterstock: 8 (Igor Plotnikov), 10 (Maslowski Marcin), 11 right (Kzenon), 14 (Mario Krpan), 15 (Geza Farkas), 18 top (Manuel Budijono), 18 bottom (olko1975), 19 (Balate Dorin), 20 (Richard Johnson), 24 (Steve Bridge), 25 (Huw Fairclough), 26 (Steve Heap), 29 (Peter Rhys Williams), 31 (Marina VN), 32/33 (Anton Havelaar), 34 (Andrii_K), 36/37 (Anton Havelaar), 36 (Maria_Usp), 37 top left (dmevans), 38 (Little Adventures), 39 (MaTiFo), 40 left (365 Focus Photography), 44 (Reme-jimenez), 45 left (Sopotnicki), 45 bottom right (Antonina Vlasova), 46 (K303), 48 (Yeinism), 50 (barmalini), 51 (Damian Lugowski), 53 (Igor Sibu), 57 left (Ruud Morijn Photographer), 62 (Tong_Stocker), 64/65 (Emilio JoseG), 69 (chinahbzyg), 70 (Kevin Standage), 72 (Sanatana), 73 top right (chinahbzyg), 74 (CRS Photo), 78 (redstone), 80 (Suprabhat), 81 (Sukpaiboonwat), 82 (Hari Mahidhar), 83 (T photography), 84 (Saiko3p), 85 (MPPhotograph), 87 (My Good Images), 88 (Zakariya AF), 89 bottom right (Indonesiapix), 91 (Izz Hazel), 95 right (Rich Carey), 96 (JKuy), 97 top right (Arshad DM), 98 (Tradol), 99 (AuGustz), 101 right (Nuttawut Uttamaharad), 104 (hangingpixels), 107 (Hypervision Creative), 109 (Takuya Tomimatsu), 124 (Sunshine Seeds), 125 (Victoria Field), 126 (Foto Stellanova), 127 Boulenger Xavier), 129 left (Photoagriculture), 130 (its.mwangi), 131 left (Stanley Njihia), 138 (Dewald Kirsten), 144 & 145 both (Torsten Pursche), 147 (Kaikups), 149 right (Boulenger Xavier), 152 (Tony Campbell), 154/155 (Jordan Feeg), 156 (Ramon Cliff), 162/163 (Russ Heinl), 164 (Devan Rawn), 165 (drachorn), 166 (JJ Gouin), 168 (Don Landwehrle), 169 (Sharon Lumpkin), 173 (Cynthia Baldauf), 175 (Diamond D Images), 176 (Lblackwellphotography), 178 top (Charles F Gibson), 179 (Joe Lafoon), 180 right (Anne Kitzman), 188 (photosounds), 191 top left (Naya Dadara), 191 top right & 191 bottom (Heidi Besen), 201 (Erica Ruth Neubauer), 204 (Paulo Nabas), 205 (Felix Santiago Allendes), 207 top right (ChrisW), 207 bottom right (Mark Skalny), 212 (Fernando filmmaker), 217 & 218 (Leonardo Loyola), 2319 (Wirestock Creators), 222 left (Alf Ribeiro), 223 (Jular Seesulai)

Shutterstock Editorial: 76 (Dasril Roszandi/NurPhoto